Maths And The Great Beyond

Maths And The Great Beyond

Charlotte Harris

Copyright © 2015 by Charlotte Harris.

ISBN: Softcover   978-1-5144-6134-1
           eBook   978-1-5144-6135-8

All rights reserved. No part of this book may be reproduced or transmitted in any form or by any means, electronic or mechanical, including photocopying, recording, or by any information storage and retrieval system, without permission in writing from the copyright owner.

Any people depicted in stock imagery provided by Thinkstock are models, and such images are being used for illustrative purposes only. Certain stock imagery © Thinkstock.

Print information available on the last page.

Rev. date: 06/25/2015

To order additional copies of this book, contact:
Xlibris
800-056-3182
www.Xlibrispublishing.co.uk
Orders@Xlibrispublishing.co.uk
688214

ONE

The year was 1800. The dawn of a new era was unfolding.

A successful brood of mathematicians were forming a

group of like-minded thinkers who were to take the world

by storm in their field and many others towards an educated

liberation and what was to become a power surge of intellect, teamwork, and logic. The resulting domino effect that

their mathematics
had on the maths
world was a
puddle of new

equations every five years or so in the great ocean of truth that began in 5 BC and swamped about at every 100-year mark, beginning not with Plato, Aristotle, and Alexander the Great, but in

1805 with Innes, Gustav, and Le Bonne.
It was an

achievement, even in those days, to have an education.

After the High Renaissance, it must have been a climb down

to find anyone who did not have an interest in music and

mathematics, but then again, times were hard and it was difficult to be taught across a range of subjects that included maths and music and fund for all of them when these were expensive and exclusive and those were more important for a career after a schooling, so many people at this time began life with no formal education at all. Those with an education in mathematics were keen to

educate the whole world, it seems, and there was an

influx of formulae and reason, which was when algebra took

root and grew,
and formulae
led to equations,
and equations

led to more formulae, and the fundamental building blocks led on.

Late in the fifties, mathematics became more intense, less

justifiable with
all the hard
work, and more
structured in the

approach. New mathematicians came to the fore with fresh ideas, breaking new ground at every turn and did everything but

stagnate. So much had been covered over the last 150 years, but then,

since Diophantus
writing his
famous $x^2 + y^2 = z^2$, no real

fundamental
advancement
had been made
on any ground

that had been set before, and especially on the particularly difficult problem of how to add numbers with powers. The

building block
approach that
provides the climb
with shapes that

fit to approximate
a wall as closely
as possible was
always the best

strategy. This strategy did not allow for improvement to

building blocks, so there was no advancement to fundamentals,

only the possibility
of better building
blocks. No one
ever thought to

improve upon our addition or multiplication skills. Addition

was addition,
subtraction was
subtraction,
multiplication

was multiplication, and division was division. Then there were other building blocks including further number concepts.

It is easy to imagine that numbers are formed from

other numbers in a form of addition, or have roots and factors,

and that they,
in some sublime
way, have a
fundamental

property of being
subtracted from,
divided into, and
being made up

of other numbers in the form of a number pair, for instance primes,

or as a number
that repeats itself
in smaller units
of groups or

even a pattern.
When we look
at numbers in
a fundamental

way, we are looking for a solution that is a whole number

or one that can be written in a concise way, such as a real

number with a fraction or root or number with a fraction or root.

Such numbers may be in fact a decimal number but are written

in such a way
as to imply that
they can be put
on paper easily

as a whole form
that can describe
the number in
its total with no

remainders left on the calculator. Pi is the decimal number 3.141592

653589793238

... but it can also be expressed as real number π

describing the number as a complete concept as if it were

a whole number. The way numbers are grouped is important, and how parts group together to form a whole. If we can understand

that the way in
which numbers
form a pattern
may be identical

to the way
in which some
other numbers
behave with the

same pattern and way in which we find the result. The answer may be remarkable too; for instance, if the similar number is

increased by a factor of three, then the answer may also be

increased by a factor of three. If we have an equation

$4^2 = (3 + 1)^2$, we can work out several things: namely that

4 = 3 + 1 and also then 4/4 = 3/4 + 1/4, but also when this is

a generalization
that applies to
all numbers,
it is more

convenient to talk in terms of x and y, especially if one or more of the numbers are not known. Then x = n forms the basis of our equation when *x* is a quantity (any large number will do,

usually a whole number greater than one, and less frequently a negative number) and n is our counting system with any number in a range starting from 1 and often 0.

TWO

The addition of exponents with like bases has

a possible solution in

$$A^x + A^y = [(A^{x-y} + 1) / A^{x-y}] \times A^x.$$

The addition of exponents with like powers has a definite solution in $G^x + A^x = [(G/A)^x + 1] \times A^x$. This solution reflects well the fundamental law of addition to multiplication. The subtraction of exponents with like bases has a possible solution in

$A^x - A^y = [(A^{x-y} - 1) / A^{x-y}] \times A^x$.

The subtraction of exponents with like powers has a definite solution in

$G^x - A^x = [(G/A)^x - 1] \times A^x$.

This shows that adding or subtracting exponents

produces a real number answer similar to A^x.

A general solution

A^m for any whole number power *m* can be found in adding and subtracting exponents using A^m as the multiplier. Then

m is more usual in x > m > y but otherwise still holds.

The addition of exponents in any other exponent is

$A^x + A^y = [A^m + (1 / A^{x-y-m})] \times A^{x-m}$.

The addition of exponents can be expressed as well

$$A^x + A^y = [A^{x-m} + (1/A^{m-y})] \times A^m.$$

The subtraction of exponents in any other exponent is

$$A^x - A^y = [A^m - 1 + (A^{x-m-y} - 1) / (A^{x-m-y})] \times A^{x-m}.$$

Then a better answer for m = n is

$$A^x - A^y = [(A^{x-n} - 1) + ((A^{n-y} - 1) / A^{n-y})] \times A^n.$$

www.ingramcontent.com/pod-product-compliance
Lightning Source LLC
Chambersburg PA
CBHW030909180526
45163CB00004B/1769